SPOTLIGHT ON PHYSICAL SCIENCE

HEAT

IT'S ENERGETIC

JILL KEPPELER

PowerKiDS press™

NEW YORK

Published in 2020 by The Rosen Publishing Group, Inc.
29 East 21st Street, New York, NY 10010

Editor: Jennifer Lombardo
Book Design: Michael Flynn

Photo Credits: Cover, p. 1 Marian Weyo/Shutterstock.com; (series molecular background) pro500/Shutterstock.com; p. 4 Africa Studio/Shutterstock.com; p. 5 Sergey Novikov/Shutterstock.com; p. 6 Smileus/Shutterstock.com; p. 7 Monkey Business Images/Shutterstock.com; p. 9 Joshua Resnick/Shutterstock.com; p. 10 LutsenkoLarissa/Shutterstock.com; p. 11 nadianb/Shutterstock.com; p. 13 Daniel Jedzura/Shutterstock.com; p. 14 HappyRichStudio/Shutterstock.com; p. 15 Designua/Shutterstock.com; p. 17 anyaivanova/Shutterstock.com; p. 19 Dietrich Leppert/Shutterstock.com; p. 20 Nikki Zalewski/Shutterstock.com; p. 21 Soloviova Liudmyla/Shutterstock.com; p. 23 Imagno/Hulton Fine Art Collection/Getty Images; p. 24 biDaala_studio/Shutterstock.com; p. 25 Hulton Deutsch/Corbis Historical/Getty Images; p. 27 UniversalImagesGroup/Getty Images; p. 28 Fotos593/Shutterstock.com; p. 29 Antonio Guillem/Shutterstock.com; p. 30 AlexMaster/Shutterstock.com

Cataloging-in-Publication Data

Names: Keppeler, Jill.
Title: Heat: it's energetic / Jill Keppeler.
Description: New York : PowerKids Press, 2020. | Series: Spotlight on physical science | Includes glossary and index.
Identifiers: ISBN 9781725313149 (pbk.) | ISBN 9781725313170 (library bound) | ISBN 9781725313156 (6 pack)
Subjects: LCSH: Heat--Juvenile literature. | Thermodynamics--Juvenile literature.
Classification: LCC QC256.K436 2020 | DDC 536--dc23

Manufactured in the United States of America

CPSIA Compliance Information: Batch #CWPK20. For further information contact Rosen Publishing, New York, New York at 1-800-237-9932.

CONTENTS

CHAPTER ONE

HOT, HOT, HOT!

What do you think of when someone says the word "heat?" Do you think of the sun beating down on you at the beach? What about the sight and sound of a pot of boiling water on the stove? Maybe you think of the feel of the sidewalk under your bare feet during the summer, or the taste of a piping hot bowl of soup.

Heat may seem like a simple thing—so simple that you barely need to think about. However, it's a key part of many ideas and principles of physical science. Although many people use the word "heat" to talk about how hot or cold something is, it's not that simple. In **physics**, heat is the transfer of energy from one thing to another because of a difference in temperature. Heat moves from warmer things to cooler things.

A trip to the beach can be a great way to spend a nice hot day!

CHAPTER TWO

ENERGETIC

You can't have heat without energy. In physics, energy is the ability to do work. However, "work" in this example doesn't mean just jobs or chores. It means when a force acts on an object to move it. This is true for both very small objects and very big objects!

Energy is a very basic idea of science. It can mean many different things, including the energy you have to run around the playground!

There are many kinds of energy. Potential energy is stored energy in an object. Kinetic energy is the energy of an object in motion. Chemical energy is stored in chemical compounds. There's also nuclear energy and mechanical energy. **Thermal** energy is energy that causes atoms, the smallest particle of an element that can exist, to move faster. This is also called heat energy, but it's not the same as heat or temperature. The difference can be confusing. However, heat, temperature, and thermal energy are all related.

CHAPTER THREE

WHAT'S THE DIFFERENCE?

The difference between heat, temperature, and thermal energy all starts with atoms and molecules—the tiny particles that make up all matter. These particles are always moving. Because they're always moving, they have kinetic energy.

Temperature is a measure of how hot or cold something is. It's also the measure of how fast the particles in that thing are moving. When the particles speed up, the thing gets warmer. When they slow down, the thing gets colder.

Thermal energy, however, is the amount of kinetic energy all of a thing's particles have in total. Something that's hot—such as a steak being cooked on a grill—will have more thermal energy that something cold—such as a frozen steak just removed from the freezer. The particles within the grilled steak are moving faster than the particles in the frozen steak.

This steak has a lot of thermal energy, which means its particles are moving very quickly—even though the steak stays still!

Temperature isn't affected by how big an object is, but thermal energy is. A larger object will have more thermal energy than a smaller object that's the same temperature. For example, a big steak hot from the grill may have the same temperature as a small sliver of meat cooked at the same time, but it will have more thermal energy. A large cold object may have more thermal energy than a very small warm object, even though its temperature is lower, just because there are more total particles in it to move.

Heat fits in with all of this because it's the transfer of this thermal energy from a warmer object to a colder one. Thermal energy is, in a way, the amount of heat something could give off.

The big piece of steak to the left and the smaller slices of steak to the right may have the same temperature, but the big piece would have more thermal energy.

CHAPTER FOUR

GO WITH THE FLOW

Thermal energy always tries to balance itself out. If you come inside after shoveling snow and wrap your hands around a mug full of hot cocoa, thermal energy will move from the mug to your hands. That's heat! Your hands will get warmer, but the mug will get cooler. If your hands were hot—say, after playing basketball on a warm summer day—and you wrapped them around something cold, such as a glass of iced lemonade, the opposite would happen. Your hands would cool down, while the glass would warm up.

Heat always flows from warmer things to cooler things. It doesn't work the other way around. It's almost impossible to **confine** thermal energy to one thing, too. Think about trying to keep food warm after it's cooked. That can be tough! The energy wants to flow.

Hot cocoa warms you up because its thermal energy flows to you. (Plus, it tastes good!)

CHAPTER FIVE

STATES OF MATTER

When a cooler object is in contact with a warmer one, the flow of thermal energy will continue until the objects reach a state of thermal **equilibrium**. This means the temperatures have balanced out and are about the same.

Sometimes, though, something can absorb heat without its temperature immediately going up—if it changes from one physical **phase** to another. The main states, or phases, of matter are solids, liquids, and gases.

A change in state happens when the particles in an object rearrange, breaking bonds, in reaction to a transfer of energy instead of just speeding up. In general, an increase to a certain temperature will turn solids into liquids and liquids into gases. The reverse is also true. Each substance has a different temperature point at which it will change phases.

STATES OF MATTER

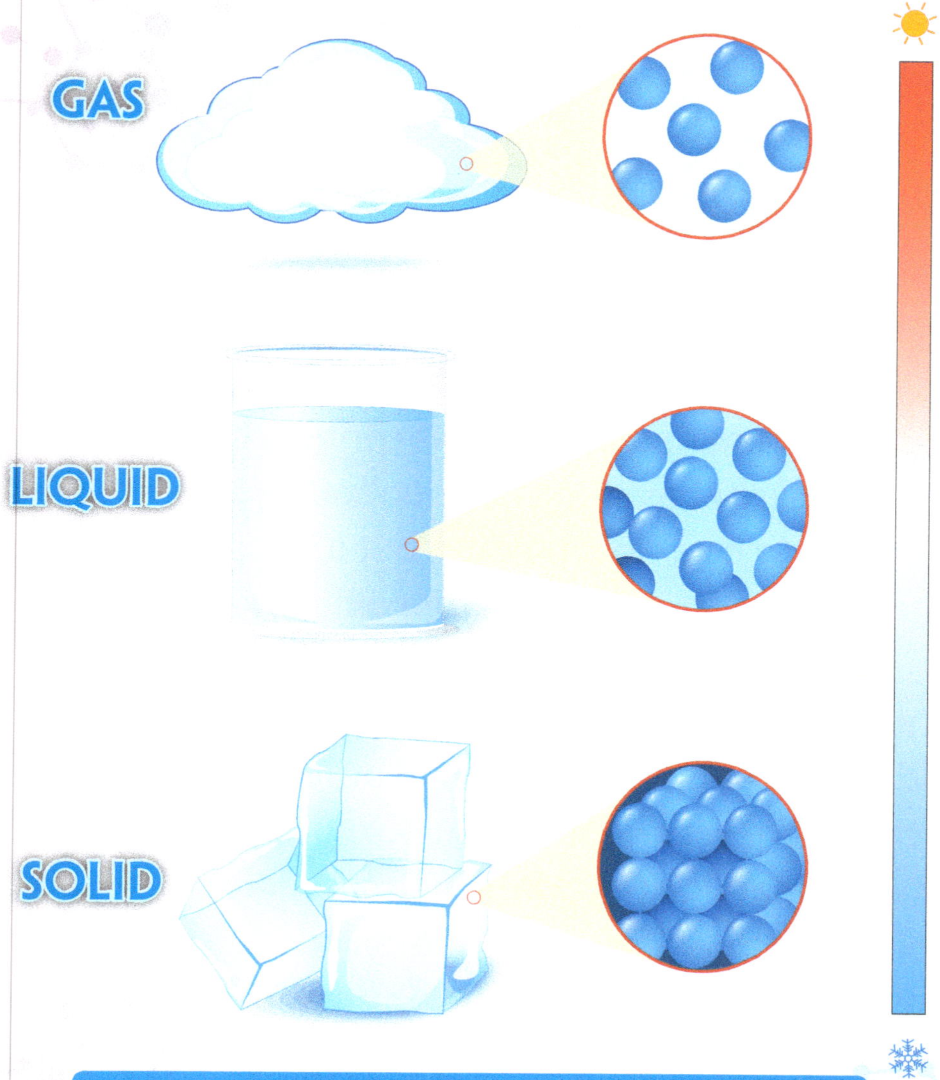

The molecules of a solid, such as ice, are more tightly packed than the molecules of a liquid, such as water. The molecules of a liquid are more tightly packed than those of a gas.

CHAPTER SIX

CONDUCTION

So, heat is the transfer of thermal energy. There are three ways this transfer can happen. These ways are called conduction, convection, and radiation.

Conduction happens when two objects with different temperatures are touching. The particles of one object **collide** with the particles of the other object, transferring energy directly. This continues until the thermal energy has spread throughout the other object. This tends to be a slow process, but it happens faster in materials with large temperature differences.

Solids conduct heat better than liquids or gases because their particles are closer together, and some solids are better conductors than others. Metals in particular are very good conductors of thermal energy. Materials that don't conduct thermal energy well, such as plastic, rubber, and wood, are called insulators. They slow the flow of energy. That's why many cooking pans have plastic handles.

What would happen if you picked up a metal pan that had been heated on top of a stove? You might burn yourself! Energy spreads from the stove to the pan and up the metal handle.

CHAPTER SEVEN

CONVECTION

While conduction happens mostly with solids, convection occurs with liquids and gases. Both of these states of matter can flow. In convection, warmer parts of a liquid or gas move and take the place of cooler parts of that liquid or gas.

This has to do with **density**. When a gas or liquid gets hotter, it expands and becomes less dense. This means its particles spread out and take up more space. Because of this, warmer gases or liquids rise, while cooler gases or liquids fall. Then the warmer substances cool down—and the cooler ones warm up. The process repeats itself, setting up a convection current.

Think of a heater in a cold room. As the air near the heater warms, it rises. It pushes aside colder air, which falls back down to the warmer areas, where that air heats and rises. In this way, the air keeps **circulating** through the room.

Have you ever seen a lava lamp? These neat lamps work because of convection. A lava lamp contains wax that's melted by a light bulb at the bottom of the lamp. The lighter, less dense melted wax floats to the top, cools, and sinks back down—again and again.

CHAPTER EIGHT

RADIATION

While conduction and convection both need particles to be affected by thermal energy, heat transfer by radiation is different. It doesn't need particles or contact between a heat source and an object or substance. Radiation is waves of energy, such as those **emitted** by the sun or any other hot object. These waves can move through space or even some solid materials until they hit something and are either absorbed or reflected. They travel at the speed of light!

Think of holding your hands out toward a campfire. Feel the heat? That's radiation.

Thermal energy is carried by infrared radiation, a type of **electromagnetic** radiation, and energy transferred this way is called radiant energy. Some substances absorb or reflect it better than others. For example, if you wear a black shirt on a sunny day, you might feel warmer than if you wear a white shirt. Black surfaces absorb radiant energy better than paler surfaces.

CHAPTER NINE

MEASURING HEAT AND TEMPERATURE

Because heat and temperature are two different things, they have to be measured two different ways. We measure temperature with **thermometers** and temperature scales. There are three scales that people often use: Fahrenheit, Celsius, and Kelvin. People mostly use the Fahrenheit scale in the United States. This is the scale you'll probably see if you look at a thermometer at your house. Scientists and people in most other countries use the Celsius scale. Scientists also sometimes use the Kelvin scale.

Thermometers existed long before these temperature scales, however. Italian scientist Galileo Galilei invented the first simple thermometer near the end of the 1500s. Scientist Daniel Gabriel Fahrenheit created the first modern sealed thermometer—and the temperature scale that was named after him—in the early 1700s. On this scale, water freezes at 32°Fahrenheit and boils at 212°Fahrenheit.

Galileo was, in many ways, the founder of modern physics. He lived from 1564 to 1642.

Heat, on the other hand, is measured in units called joules or calories. It can also be measured in British thermal units (BTUs). A BTU is how much heat is needed to raise the temperature of a pound of water by one degree Fahrenheit. A calorie is how much heat is needed to raise the temperature of a gram of water by one degree Celsius.

A joule is used to refer to one unit of work or energy, so it applies to more than just heat. It's named after English scientist James Prescott Joule, who did many experiments involving heat and energy in the 1800s. In fact, Joule's work led to the basis of a very important principle of modern science—the law of conservation of energy, which is the first law of **thermodynamics**.

James Prescott Joule, shown here, figured out that different forms of energy are the same—and that they can be turned into each other.

CHAPTER TEN

THE HISTORY OF HEAT

Joule was an important player in the study of heat, but he wasn't alone. Other scientists did a lot of experiments studying this important principle of energy. Galileo, Robert Boyle, and Isaac Newton all studied heat.

Scottish scientist Joseph Black studied heat during the 1700s. He figured out that heat and temperature weren't the same thing. He also came up with the idea of latent heat. This is the energy let off or taken in by something when it changes its state of matter but doesn't change its temperature. French scientist Antoine-Laurent Lavoisier studied heat at about the same time, as well as chemical reactions.

By the end of the 1700s, scientist Benjamin Thompson (also known as Count von Rumford) had figured out the part kinetic energy played in heat. This was the start of the field of thermodynamics.

Joseph Black, shown here, did many experiments with heat. He was friends with American Benjamin Franklin.

CHAPTER ELEVEN

THREE LAWS

The three laws of thermodynamics are very important to science. The first law says that the total amount of energy (including heat) in a system and its surroundings doesn't change. The energy is conserved. Energy can't be created nor destroyed, so the total energy of the universe is constant. Energy may, however, change forms.

No matter how cold you think it is outside, it can't reach absolute zero! The laws of thermodynamics are **complex**, but the basic ideas are very important.

The second law of thermodynamics says that not all the heat in a system can be changed to work or another form of energy. Some will always remain heat. This is partially because no particular system can become less disordered, and heat is a very disordered kind of energy.

The third law of thermodynamics says that no system can reach absolute zero, which is the coldest temperature. While science has the idea of absolute zero, a system can't actually reach it.

CHAPTER TWELVE

THE HEAT IS ON

Heat is a such a simple fact of life that you probably don't think about it much unless you're very warm or very cold. Still, this scientific principle is very important to the study of energy and how we can use it. Scientists continue to study and learn more about heat even today.

The next time you sip a warm cup of hot cocoa or walk on hot sand at the beach, think about heat and the difference between it, thermal energy, and temperature. When you throw a snowball at a friend or take a bite of a melting popsicle, think about the states of matter and how things change. When you watch soup cooking on the stove or hold your hands over a campfire, consider conduction, convection, and radiation. Science is an everyday part of life!

GLOSSARY

circulate (SUHR-kyuh-layt) To move in a circular path without stopping through a system or place.

collide (kuh-LYD) To crash into each other.

complex (kahm-PLEHKS) Not easy to understand or explain; having many parts.

confine (kuhn-FYN) To keep indoors or within other limits.

density (DEHN-suh-tee) The amount of matter in a given area.

electromagnetic (ih-lehk-troh-mag-NEH-tik) Relating to the magnetic field that's produced by a current of electricity.

emit (EE-miht) To give off.

equilibrium (ee-kwuh-LIH-bree-uhm) A state in which things and forces are balanced.

phase (FAYZ) A stage in a process or sequence of events.

physics (FIH-ziks) The study of matter, energy, force, and motion, and the relationship among them.

thermal (THUHR-muhl) Having to do with heat.

thermodynamics (thuhr-moh-dy-NAA-miks) A science that deals with heat and related forms of energy.

thermometer (thuhr-MAH-muh-tuhr) A tool used for measuring temperature.

INDEX

PRIMARY SOURCE LIST

Page 23
Galileo Galilei. Oil on canvas portrait. 1636. Hulton Fine Art Collection.

Page 25
James Prescott Joule. Illustration. c. 1870. Wikimedia Commons.

Page 27
Joseph Black. Illustration. c. 1750. Günther Bugge: *Das Buch Der Grossen Chemiker*. Wikimedia Commons.

WEBSITES

Due to the changing nature of Internet links, PowerKids Press has developed an online list of websites related to the subject of this book. This site is updated regularly. Please use this link to access the list: www.powerkidslinks.com/SOPS/heat